YOUR KNOWLEDGE HAS VALUE

Islam Ud Din Khan

Antimicrobial susceptibility patterns and proportions of Escherichia coli in urinary tract infections in Mansehra, Pakistan

GRIN Verlag

Bibliografische Information der Deutschen Nationalbibliothek:

Die Deutsche Bibliothek verzeichnet diese Publikation in der Deutschen National-
bibliografie; detaillierte bibliografische Daten sind im Internet über http://dnb.d-
nb.de/ abrufbar.

Imprint:

Copyright © 2013 GRIN Verlag GmbH
Druck und Bindung: Books on Demand GmbH, Norderstedt Germany
ISBN: 978-3-656-57510-8

This book at GRIN:

http://www.grin.com/en/e-book/265002/antimicrobial-susceptibility-patterns-and-
proportions-of-escherichia-coli

GRIN - Your knowledge has value

Der GRIN Verlag publiziert seit 1998 wissenschaftliche Arbeiten von Studenten, Hochschullehrern und anderen Akademikern als eBook und gedrucktes Buch. Die Verlagswebsite www.grin.com ist die ideale Plattform zur Veröffentlichung von Hausarbeiten, Abschlussarbeiten, wissenschaftlichen Aufsätzen, Dissertationen und Fachbüchern.

Visit us on the internet:

http://www.grin.com/

http://www.facebook.com/grincom

http://www.twitter.com/grin_com

Antimicrobial susceptibility patterns and proportions of *Escherichia coli* in urinary tract infections in Mansehra, Pakistan

Islam Ud Din Khan

Department of Microbiology, Hazara University Mansehra, Pakistan

Urinary tract infections are the most common bacterial infections globally, caused by *Escherichia coli*. *Escherichia coli* produces an enzyme called extended spectrum β-lactamases (ESBL) which inhibits penicillins, cephalosporins and various other antibiotics. The current study included 1720 specimens, isolated from urine samples of inpatients and outpatients suffering from Urinary tract infections. The antimicrobial susceptibility by disc diffusion was performed on each isolate by using 10 antibiotics according to Clinical Laboratory Standards Institute (CLSI) criteria. 370 (21.5%) specimens were confirmed to be *E.coli* isolates. *E.coli* isolates were found to be 97.2% sensitive against Imipenem, 96.4% against Meropenem, 50.0% against Gentamicin, 47.2% against Kanamycin, 38.3% against Ciprofloxacin, 15.6% against Doxycycline, and 25.6% sensitive against Co-trimoxazole. A large proportion of *E.coli* isolates were found to be multi drug resistant. *E.coli* isolates were found to be 91.8% resistant against Ampicillin, 84.3% against Doxycycline, 82.4% against Cefaclor, and 80.5% resistant against Nalidixic Acid.

Key words: Antimicrobial resistance, *Escherichia coli*, Extended Spectrum β-Lactamase, Urinary Tract Infections.

1

Introduction

Escherichia coli is found in the human gut, and is the common pathogen causing urinary tract infections (Wagenlehner et al., 2008; De Francesco et al., 2007; Kashef et al., 2010). Antimicrobial resistance in *E. coli* has been reported worldwide and the increasing rates of resistance among *E. coli* are a growing concern in both developed and developing countries (Bell et al., 2002; El Kholy et al., 2003). Antimicrobial resistance makes infections complicated to be treated. Generally, almost 95 % cases with severe symptoms are treated without bacteriological investigation (Dromigny et al., 2005). Occurrence and susceptibility profiles of *E. coli* show substantial geographic variations as well as significant differences in various populations and environments (Erb et al., 2007). In Pakistan, a number of studies have been done on the prevalence of *E.coli* and its antimicrobial susceptibility patterns. The aim of this study was to determine the proportions of *E.coli,* and its antimicrobial susceptibility patterns for efficient management of urinary tract infections.

Material and methods

Bacterial isolates

In this study, *E.coli* isolates were collected from public and private hospitals and healthcare centers from June 2009 to March 2010. Only one sample was obtained from each patient and included in the study. Clean catch midstream urine specimens were collected in sterilized bottles and submitted to clinical microbiology laboratory. The samples received were inoculated onto Blood agar and McConkey agar, and then incubated in aerobic atmosphere at 37ᵒC for 24 hours. *E.coli* isolates were confirmed by biochemical tests. The plates yielding growth as per Kass counts (single specie count more than 105 organisms/ml) were processed further (Kass, 1956).

Antimicrobial susceptibility testing

Antimicrobial susceptibility testing was performed by the modified Kirby Bauer disc diffusion method on Muller-Hinton agar (Oxoid, England) as described by the Clinical Laboratory Standard Institute (CLSI) (CLSI, 2006).

The 10 antimicrobial agents used were; Meropenem, Imipenem, Gentamicin, Kanamycin, Ciprofloxacin, Doxycycline, Co-trimoxazole, Ampicillin, Cefaclor and Nalidixic Acid. *E. coli* NCTC 10418 was used for quality control for antimicrobial susceptibility tests.

Statistical analysis

Statistical analysis was performed by the chi-square test and *P* values of ≤0.05 were considered significant.

Results

A total of 1720 samples were analyzed for isolation and identification of bacteria and antimicrobial susceptibility testing from June 2009 to March 2010. *E. coli* was isolated from 370 (21.5%) samples.

Age range of patients was in between 1 year to 80 years. More isolates were recovered from female patients as compared to male patients the ratio being 3:1. Middle aged patients i.e. 20 - 43 years accounted for 54.3% of infections while, the second predominant group that was 1 – 9 years accounted for 24% infections.

Among the β-lactam antibiotics, the most effective was Imipenem with 97.2% of the isolates susceptible to this agent, followed by Meropenem with 96.4% of the isolates being susceptible. The antimicrobial susceptibility of *E.coli* against Gentamicin was 50.0%, Kanamycin 47.2%, Ciprofloxacin 38.3%, Doxycycline 15.6%, and Co-trimoxazole 25.6%. The highest resistance was found to be 91.8% against Ampicillin, followed by 84.3% against Doxycycline. The antimicrobial resistance against Cefaclor was 82.4% and, Nalidixic Acid 80.5%. The antimicrobial susceptibility testing results against the three antibiotics-- Kanamycin, Co-trimoxazole and Cefaclor were partial. The antimicrobial susceptibility for the 10 antibiotics used in this study is mentioned in Table 1.

Table 1._ Antimicrobial susceptibility patterns of *E.coli*

NAME OF ANTIBIOTIC	RESISTANT	(%)	SENSITIVE	(%)	INTERMEDIATE (%)	
IMIPENEM	10	2.7	360	97.2	0	0
MEROPENEM	13	3.5	357	96.4	0	0
GENTAMICIN	185	50.0	185	50.0	0	0
KANAMYCIN	188	50.8	174	47.2	8	2.1
CIPROFLOXACIN	228	61.6	142	38.3	0	0
DOXYCYCLINE	312	84.3	58	15.6	0	0
CO-TRIMOXAZOLE	270	72.9	95	25.6	5	1.3
AMPICILLINE	340	91.8	30	8.1	0	0
CEFACLOR	305	82.4	50	13.5	15	4.05
NALIDIXIC ACID	298	80.5	72	19.4	0	0

Discussion

This study reveals the proportions and antimicrobial resistance of *E.coli* in patients suffering from urinary tract infections, in Mansehra, Pakistan. Majority of *E. coli* were isolated from female patients, (77.0%) similar to other reports (El Astal, 2005; Hasan et al., 2007). Patients from adult age group (22 - 45 years) bestowed 54.3% isolates, which is similar to another study (Akram et al., 2007).

Among the β-lactams antibiotics tested, the Carbapenems have the widest spectrum of activity. Imipenem was the most effective antibiotic to which 97.2% isolates were susceptible. However, a study in India has reported 100% activity for Imipenem against *E. coli* (Akram et al., 2007). The antimicrobial susceptibility to Meropenem in this study was 96.4%. In Poland, the antimicrobial susceptibility to Meropenem was reported 100% (Hryniewicz et al., 2001).

Resistance to Ampicillin in *E. coli* has been reported very high in Pakistan (Rahman et al., 2002; Zaidi et al., 2005). In the present study, 91.8% resistance against Ampicillin was found which is similar to that reported in Jordan (Shehabi et al., 2004). Noor et al. (2004) have reported 100% resistance to Ampicillin in *E. coli* from Pakistan.

The antimicrobial resistance to Gentamicin was found 50.0% in this study, while in Israel and India, it was reported 71% and 64% respectively (Tankhiwale et al., 2004; Colodner et al.,

2007). The high resistance to Gentamicin in Israel and India is may be due to the increased usage of this drug.

In this study, 50.8% isolates of *E.coli* were resistant to Kanamycin while, in Karachi, Pakistan, it was reported 50.0% which is lower than the present study (Gul et al., 2004).

Ciprofloxacin has been recommended as a first line therapy in urinary tract infections (Paterson, 2000). But, susceptibility to flouroquinolones is decreasing throughout the world (Matute et al., 2004; El Astal, 2005; Karlowsky et al., 2006). The resistance to Ciprofloxacin was 61.6% in this study, which is higher than reported in other parts of the world like Palestine, Canada, USA and Turkey (Mazzulli et al., 2001; Sahm et al., 2001; El Astal, 2005; Yuksel et al., 2006).

Hospital acquired pathogens are generally resistant to multiple antibiotics due to the increased usage of these antibiotics. Surveillance studies have been conducted to monitor antimicrobial susceptibility patterns in pathogenic bacteria to help clinicians when using empirical treatment for infections. Studies from USA, Europe and many other countries have shown better susceptibility patterns for pathogens isolated from urinary tract infections against Co-trimoxazole (Szczypa et al., 2001; Mazzulli et al., 2001; Sahm et al., 2001; Gordon and Jones, 2003; Bonsu et al., 2006). But, *E.coli* has also shown greater resistance to Co-trimoxazole in these countries (Tankhiwale et al., 2004; Akram et al., 2007). A reason for this increasing resistance may be due to the extensive use of Co-trimoxazole. In the present study, the antimicrobial resistance to Co-trimoxazole was 72.9%. Hence, Co-trimoxazole cannot be recommended as an empiric treatment for the treatment of UTIs in Pakistan.

This study shows that a large number of *E.coli* isolates are resistant to many antibiotics, such as Ampicillin, Doxycycline, Cefaclor and Nalidixic Acid, which are important drugs in the treatment of urinary tract infections. Such isolates are also resistant to Co-trimoxazole and Ciprofloxacin. Carbapenems are the drugs of choice against UTIs caused by *E. coli*. Strict antibiotic policy should be adopted in hospitals to reduce the antimicrobial resistance. The antimicrobial resistance can be reduced by the decreased use of antibiotics, use of synergistic combinations, addition of an anti-resistance factor, improving the hygienic measures and regular surveillance studies (Hankook, 1998; Huovinen, 1998).

Acknowledgments

I am thankful to all staff of the department of Microbiology of Hazara University Mansehra, Pakistan for funding and proper documentation of this work.

References

Akram., M. Shahid., M. Khan., A.U. (2007). Etiology and antibiotic resistance patterns of community-acquired urinary tract infections in J N M C Hospital Aligarh, India. Ann. Clin. Microbiol. Antimicrob. 6: 4.

Bell., J.M. Turnidge., J.D. Gales., A.C. Pfaller., M. Jones., R.N. Sentry APAC Study Group (2002). Prevalence of extended spectrum beta-lactamase (ESBL)-producing clinical isolates in the Asia-Pacific region and South Africa: regional results from SENTRY Antimicrobial Surveillance Program (1998–99). *Diagn Microbiol Infect Dis*. 42:193–198.

CLSI (2006). Clinical and Laboratory Standards Institute (CLSI): Performance Standard for Antimicrobial Susceptibility Testing. 16th Informational supplement. CLSI document M100-S16.

Colodner., R. Samra., Z. Keller., N. Sprecher., H. Block., C. Peled., N. Lazarovitch., T. Bardenstein., R. Schwartz-Harari., O. Carmeli., Y. (2007). First national surveillance of susceptibility of extended-spectrum β-lactamase- producing *Escherichia coli* and *Klebsiella* spp. To antimicrobials in Israel. Diagn. Microbiol. Infect Dis. 57(2): 201-205.

De Francesco., M.A. Giuseppe., R. Laura., P. Riccardo., N. Nin., M. (2007) Urinary tract infections in Brescia, Italy: Etiology of uropathogens and antimicrobial resistance of common *Uropathogens Med Sci Moni*. 13(6): 136-144.

Dromigny., J.A. Nabeth., P. Juergens-Behr., A. Perrier-Gros-Claude., J.D. (2005). Risk factors for antibiotic resistant *Escherichia coli* isolated from community-acquired urinary tract infections in Dakar, Senegal. *J Antimicrobial Chemother*. 56: 236-239.

El Astal., Z. (2005). Increasing ciprofloxacin resistance among prevalent urinary tract bacterial isolates in Gaza Strip, Palestine. J. Biomed Biotechnol. (3): 238-241.

El Kholy., A. Baseem., H. Hall., G. Procop., G.W. Longworth., D.L. (2003). Antimicrobial resistance in Cairo, Egypt 1999–2000: a survey of five hospitals. *J Antimicrob Chemother*. 51: 625-630.

Erb., A, Stürmer., T. Marre., R. Brenner., H. (2007). Prevalence of antibiotic resistance in *Escherichia coli*: overview of geographical, temporal, and methodological variations. *Eur J Clin Microbial Infect Dis*. 26: 83–90.

Gul., N. Mujahid., T.Y. Ahmed., S. (2004). Isolation, Identification and Antibiotic Resistance Profile of Indigenous Bacterial Isolates from Urinary Tract Infections Patients. Pak. J. Biological Sci. 7(12): 2051- 2054.

Hankook., R.E.W. (1998). Resistance Mechanisms in *Pseudomonas aeruginosa* and other nonfermentative Gram-Negative Bacteria. Clin. Infect Dis. 27: S93-99.

Hasan., A.S. Nair., D. Kaur., J. Baweja., G. Deb., M. Aggarwal., P. (2007). Resistance patterns of urinary isolates in a tertiary Indian hospital. J. Ayub. Med. Coll. Abbottabad. 19(1): 39-41.

Hryniewicz., K. Szczypa., K. Sulikowska., A. Jankowski., K. Betlejewska., K. Hryniewicz., W. (2001). Antibiotic susceptibility of bacterial strains isolated from urinary tract infections in Poland. J. Antimicrob. Chemother. 47(6): 773-780.

Huovinen., P. (1998). Control of antimicrobial resistance: time for action. British Med. J. 317: 613-614.

Karlowsky., J.A. Hoban., D.J. Decorby., M.R. Laing., N.M. Zhanel., G.G. (2006). Fluoroquinolone -resistant urinary isolates of *Escherichia coli* from outpatients are frequently multidrug resistant: results from the North American Urinary Tract Infection Collaborative Alliance-Quinolone Resistance study. Antimicrob. Agents Chemother. 50(6): 2251-2254.

Kashef., N. Djavid., G.E. Shahbazi., S. (2010). Antimicrobial susceptibility patterns of community-acquired uropathogens in Tehran, Iran. *J Infect Dev Ctries*. 4(4): 202-206.

Kass., E.H. (1956). Asymptomatic infections of the urinary tract. Trans. Assoc. Am. Physicians 69: 56-64.

Matute., A.J. Hak., E. Schurink., C.A. McArthur., A. Alonso., E. Paniagua., M. Van Asbeck., E. Roskott., A.M. Froeling., F. Rozenberg-Arska., M. Hoepelman., I.M. (2004). Resistance of uropathogens in symptomatic urinary tract infections in Leon, Nicaragua. Int. J. Antimicrob. Agents 23(5): 506-509.

Mazzulli., T. Skulnick., M. Small., G. Marshall., W. Hoban., D.J. Zhanel., G.G. Finn., S. Low., D.E. (2001). Susceptibility of community Gram-negative urinary tract isolates to mecillinam and other oral agents. Can. J. Infect Dis. 12(5): 289-292.

Noor., N. Ajaz., M. Rasool., S.A. Pirzada., Z.A. (2004). Urinary tract infections associated with multidrug resistant enteric bacilli: characterization and genetical studies. Pak. J. Pharm. Sci. 17(2): 115-123.

Paterson., D.L. (2000). Recommendation for treatment of severe infections caused by Enterobacteriaceae producing extendedspectrum β-lactamases (ESBLs). Clin. Microbiol. Infect 6(9): 460-463.

Rahman., S. Hameed., A. Roghani., M.T. Ullah., Z. (2002). Multidrug resistant neonatal sepsis in Peshawar, Pakistan. Arch Dis. Child. Fetal. Neonatal Ed. 87(1): F52-54.

Sahm., D.F. Thornsberry., C. Mayfield., D.C. Jones., M.E. Karlowsky., J.A. (2001). Multidrug-resistant urinary tract isolates of *Escherichia coli*: prevalence and patient demographics in the United States in 2000. Antimicrob. Agents Chemother. 45(5): 1402-1406.

Tankhiwale., S.S. Jalgaonkar., S.V. Ahamad., S. Hassani., U. (2004). Evaluation of extended spectrum β-lactamase in urinary isolates. Indian J. Med. Res. 120(6): 553-556.

Wagenlehner., F.M. Naber., K.G. Weidner., W. (2008). Rational antibiotic therapy of urinary tract infections. *Med Monatsschr Pharm*. 31: 385-90.

Yuksel., S. Ozturk., B. Kavaz., A. Ozcakar., Z.B. Acar., B. Guriz., H. Aysev., D. Ekim., M. Yalcinkaya., F. (2006). Antibiotic resistance of urinary tract pathogens and evaluation of empirical treatment in Turkish children with urinary tract infections. Int. J. Antimicrob. Agents 28(5): 413-416.

Zaidi., A.K. Huskins., W.C. Thaver., D. Bhutta., Z.A. Abbas., Z. Goldmann., D.A. (2005). Hospital acquired neonatal infections in developing countries. Lancet 365(9465): 1175-1188.